TOOLS FOR CAREGIVERS

- **ATOS:** 0.9
- **GRL:** C
- **WORD COUNT:** 30
- **CURRICULUM CONNECTIONS:** animals

Skills to Teach

- **HIGH-FREQUENCY WORDS:** are, at, have, let's, look, these
- **CONTENT WORDS:** animal, claws, feet, hooves, pads, sticky, toenails, webs
- **PUNCTUATION:** exclamation points, periods
- **WORD STUDY:** compound word (*toenails*); long /a/, spelled ai (*toenails*); long /e/, spelled ee (*feet*); long /e/, spelled y (*sticky*); long /o/, spelled oe (*toenails*); multisyllable word (*animal*); /oo/, spelled oo (*hooves*)
- **TEXT TYPE:** information report

Before Reading Activities

- Read the title and give a simple statement of the main idea.
- Have students "walk" though the book and talk about what they see in the pictures.
- Introduce new vocabulary by having students predict the first letter and locate the word in the text.
- Discuss any unfamiliar concepts that are in the text.

After Reading Activities

Explain to readers that animals have different kinds of feet for different reasons. Ducks have webbed feet to help them swim. Eagles have claws to help them catch food. What other interesting animal feet can they think of? Can they say why the animal might need that kind of foot? Can readers name any kinds of animals that don't have feet?

Tadpole Books are published by Jump!, 5357 Penn Avenue South, Minneapolis, MN 55419, www.jumplibrary.com

Copyright ©2020 Jump!. International copyright reserved in all countries. No part of this book may be reproduced in any form without written permission from the publisher.

Editor: Jenna Trnka **Designer:** Molly Ballanger

Photo Credits: Tsekhmister/Shutterstock, cover; KETPACHARA YOOSUK/Shutterstock, 1; Katesalin Pagkaihang/Shutterstock, 3; Anneka/Shutterstock, 2br, 4–5; Makarova Viktoriia/Shutterstock, 2tr, 6–7; Robert Eastman/Shutterstock, 2mr, 8–9; FloridaStock/Shutterstock, 2tl, 10–11; Nadezhda V. Kulagina/Shutterstock, 2ml, 12–13; Villiers Steyn/Shutterstock, 2bl, 14–15; Erik Lam/Shutterstock, 16.

Library of Congress Cataloging-in-Publication Data
Names: Gleisner, Jenna Lee, author.
Title: Feet / by Jenna Lee Gleisner.
Description: Tadpole edition. | Minneapolis, MN: Jump!, Inc., (2020) | Series: Animal part smarts | Audience: Age 3–6. | Includes index.
Identifiers: LCCN 2018042926 (print) | LCCN 2018043970 (ebook) | ISBN 9781641286985 (ebook) | ISBN 9781641286961 (hardcover : alk. paper) ISBN 9781641286978 (paperback)
Subjects: LCSH: Foot—Juvenile literature.
Classification: LCC QL950.7 (ebook) | LCC QL950.7 .G54 2020 (print) | DDC 591.47/9—dc23
LC record available at https://lccn.loc.gov/2018042926

ANIMAL PART SMARTS

FEET

by Jenna Lee Gleisner

TABLE OF CONTENTS

Words to Know 2

Feet 3

Let's Review! 16

Index 16

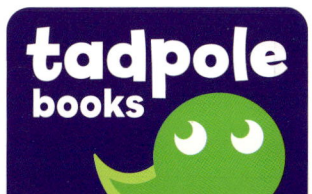

tadpole books

WORDS TO KNOW

claws

hooves

pads

sticky

toenails

webs

FEET

Let's look at animal feet!

web

These feet have webs.

6

hoof

These feet have hooves.

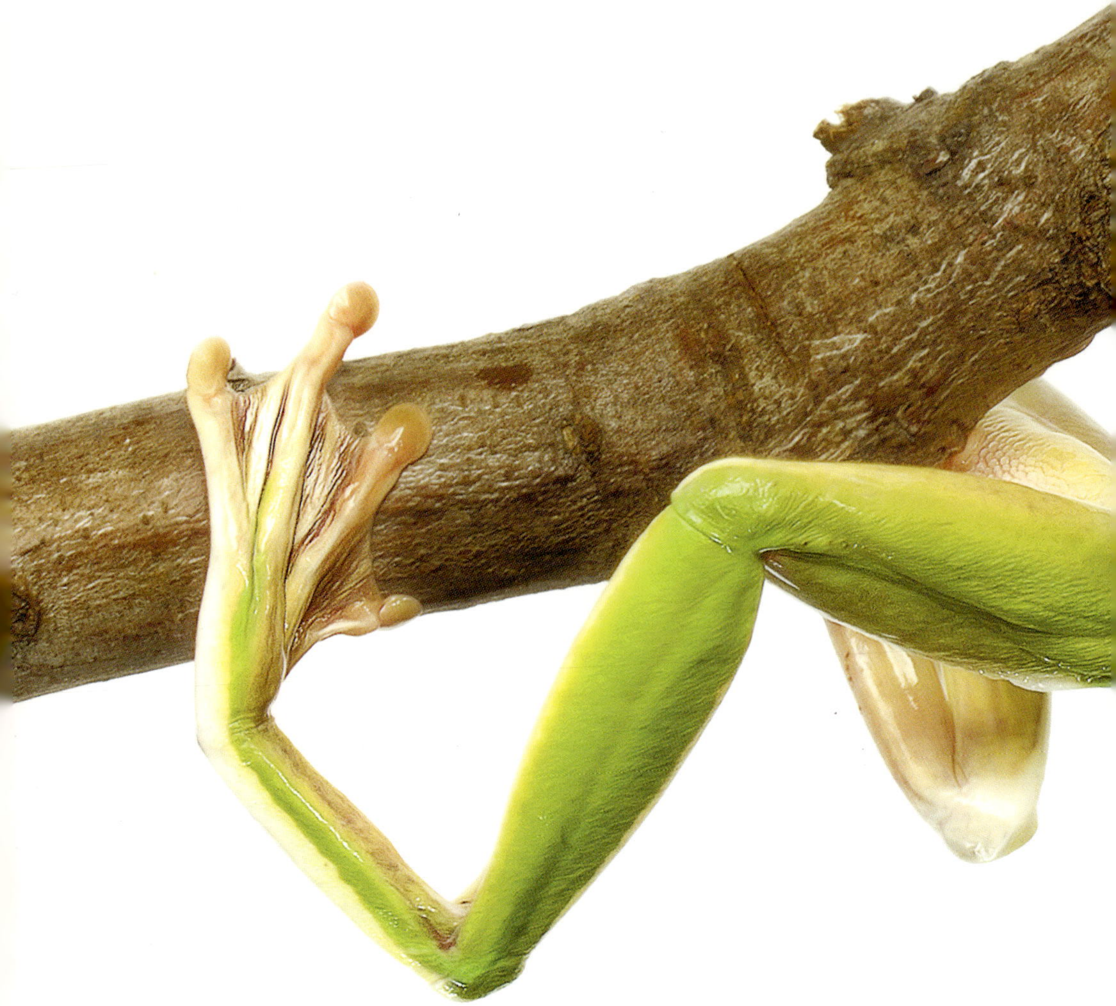

These feet are sticky.

10

claw

These feet have claws.

pad

These feet have pads.

toenail

These feet have toenails!

LET'S REVIEW!

What do these animal feet have?

INDEX

claws 11
hooves 7
pads 13
sticky 9
toenails 15
webs 5